An American Farmer

A GLIMPSE INTO AMERICA'S HERITAGE

with writings by Sue Ikerd

KPT PUBLISHING

Four generations of my husband's family have lived on our farm, and now we are blessed to see a fifth generation learn the ways of farming. My husband was born here and knows no other way of life. It's a wonderful place to raise children.

I came as a young bride and never lived so far away from a town, but I found that I loved the seclusion and beauty this place had to offer. I learned to drive a tractor, rake, and bale hay—feed silage. I watched my husband birth calves.

Our farm borders the river, a place where we loved to fish, swim, and boat. After milking in the summer, we would go to a gravel bar, where the kids would have fun in the water, and we roasted hot dogs and marshmallows on an open fire. Black walnut trees grow on the hillside along with dogwood and red bud trees with cows and baby calves grazing beneath. In the summer, I loved to pick wild blackberries and gooseberries for cobblers. Such good memories.

There were hard times. Sometimes a farmer has to battle "Mother Nature," and she can be daunting, but with God's guidance we made it through.

I know of no other place I would rather live than our family farm.

Sue Ikerd

What I learned

growing up on the farm

was a way of life

that was centered on hard work

and on faith and on thrift.

Those values have stuck with me

my whole life.

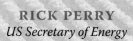

RICK PERRY
US Secretary of Energy

As a young bride, I was forewarned by my farmer husband that times would be tough on the farm because situations would arise that would require immediate attention. We never knew when equipment would break (parts and maintenance were expensive), and of course animals get sick and vets have to be called. It seemed these things happened when we were least expecting.

We also had to buy feed, fertilizer, and other regular expenses. There was very little extra money for fun things. Farm expenses would have to come first, but I believed living in the middle of this beautiful country with him would be worth what I might have to live without.

One of my favorite memories was my sun-tanned husband pulling up in the driveway of our home on his red tractor after a day of plowing the fields. With a big grin on his face, he handed me a bunch of wild daises he plucked. Still attached was a clump of plowed earth. He knew how I loved daisies. They were more beautiful to me than a dozen long stem red roses.

It's a memory

I still treasure.

The Farmer
by Sue Ikerd

He has been a farmer all of his life,

long before he took a wife.

He knew he was meant to work the soil.

His days on this earth

would be spent in toil,

Planting the crops and clearing the land.

This was all part of the Master's Plan.

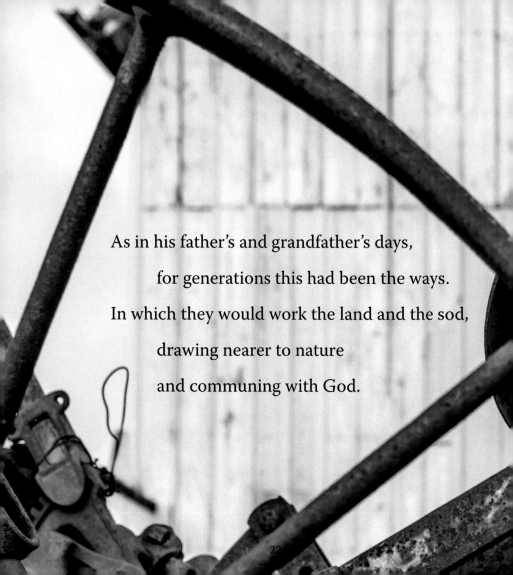

As in his father's and grandfather's days,

for generations this had been the ways.

In which they would work the land and the sod,

drawing nearer to nature

and communing with God.

To each of his neighbors he lent a hand,

They worked together to farm the land,

in autumn when the harvest came,

each one in turn did the same.

All through the week they labored each day,

but on the Sabbath they gathered to pray.

To thank Him for His blessings and love,

what they gathered on earth

had come from above . . .

When his children were born he watched them grow.

He taught them the lessons

so they would know,

and learn the ways of country and farm,

of love, truth, respect, and to do no harm

to creature on land or those in the air,

and to be good stewards of the land in their care.

He watched them ride horses and float down the stream,

but he knew that their future

could not be his dream.

This farmer he realizes he has wealth beyond measure,

because here on this farm

he has found all his treasure.

With his family around him,

for wealth there's no need.

With all of His blessings

he's a rich man indeed.

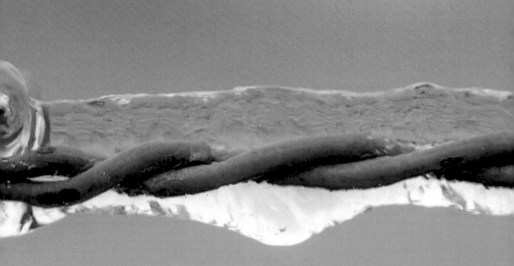

His breed is a rare one, it's becoming extinct,

with this world's busy lifestyle,

there's no time to think.

Life's becoming too hectic, and people miss out

on all of the beauty that lies roundabout.

This farmer can see it as he goes through his days,

from bird's nests to sunsets,

each free for the gaze.

The path that he's taken is different from most.

He's content in his heart and has no need to boast.

His drumbeat is different but he follows its sounds,

with his dog by his side

he walks over this ground,

of the land that he loves, he will do it no harm,

the place of his birth…

the old family farm.

"Georgie" the Bull

A few years ago my husband decided to purchase a bull to expand our beef herd. He read about Scottish Highlanders and was hoping to find a suitable one. They are a small breed, and he felt they fit our needs on the farm. To my surprise, I also found Queen Elizabeth keeps a large herd of Scottish Highland cattle at Balmoral Castle. I was excited about buying a new bull. Finding an ad for a Highlander, we called the owners and then headed off to check out the bull.

The owners were a nice retired couple. The man was quiet and reserved, but the woman was tall, thin, and so full of energy. We followed them to the pen where the bull was kept. She immediately hopped over the fence and approached the large animal. She pulled a brush from her pocket and began brushing his long cinnamon-colored, shaggy hair. Then she opened a package of blueberry muffins, and the bull happily started munching on them. On our farm we always told our kids that bulls were not to be made into pets; they can be very unpredictable. She urged me to come on over and told me I could pet him too. She said his name was Georgie. I declined but was charmed watching the interaction between this large muppet-like creature and this wiry little lady.

We purchased Georgie and happily headed home. The sweet lady sent us a supply of blueberry muffins for the bull, but I fed them to our old stock dog, instead. I didn't feel I should make a pet of Georgie.

Several years later, Georgie passed away. I loved watching that old bull grazing in the pasture. His large set of horns were quite impressive. So after a couple of years, we were able to retrieve the skull and horns. I didn't know what

I was going to do with them, but Georgie had always been special, and they were special to me. I had them lying on the carport when my grandson's fiancée saw them there. She mentioned that she would like to use them to decorate at their upcoming wedding. I was shocked and surprised. I later found out that was a trend, and many were using the horns and skulls of cattle for decorating. She asked me to make a garland of flowers to perch where Georgie's shaggy topknot had once been. His skull and horns proudly hung on the wall at the reception.

When we first purchased Georgie, I never expected him to have such a prominent place at our grandson's wedding, but he was quite a hit.

Farming looks mighty easy

when your plow is a pencil

and you're a thousand miles

from the corn field.

DWIGHT D. EISENHOWER
34th President of the United States

An American Farmer

© 2018 KPT Publishing, LLC
Written by Sue Ikerd

Published by KPT Publishing
Minneapolis, Minnesota 55406
www.KPTPublishing.com

ISBN 978-1-944833-29-9

Designed by AbelerDesign.com

First printing March 2018

10 9 8 7 6 5 4 3 2 1

Printed in the United States of America

ABOUT THE AUTHOR

SUE IKERD was born in Marshfield, MO. At age six, she moved with her dad, mom, and older sister to a small farm on the outskirts of town. When she married her husband, Don, he brought her to his river bottom farm in the Missouri Ozarks—the farm where he was born and raised and where they raised their children as well. They recently celebrated their fiftieth wedding anniversary with their family, including their two grandsons. Sue's poem, *The Farmer*, was her first and written as a tribute to her husband.

In addition to writing, she enjoys studying genealogy with her daughter, and after many years of research, she found farming has been the occupation of their ancestors for many generations—so farming is truly in their blood. You can read more of Sue's work on her website www.ourcountryblessings.com. ❧